Contents

Chapter One – Antarctica
Cold and Dark at the South Pole   4

Chapter Two – Argentina
Fire and Light in Tierra del Fuego   8

Chapter Three – Australia
Festivals and Art in Hobart   12

Chapter Four – South Africa
Whales and Sardines in Cape Town   16

Chapter Five – Ecuador
Incan Festivities in Ingapirca   20

Chapter Six – Indonesia
River Games in Pontianak   24

Chapter Seven – Nigeria
Sips, Snacks and Shade in Barage   28

Chapter Eight – Nepal
Momos and Chiya in Pokhara   32

Chapter Nine – Morocco
Work, School and Home in Marrakesh   36

Chapter Ten – China
Busy Days in Beijing   40

Chapter Eleven – Turkey
Cats and Mosques in Istanbul   44

Chapter Twelve – United States of America
Salmon and Sunshine in Ilwaco   48

Chapter Thirteen – United Kingdom
Sunrise at Stonehenge   52

Chapter Fourteen – Norway
Up All Night in Longyearbyen   56

Chapter One

Cold and Dark at the South Pole

Amundsen-Scott Research Station, The South Pole, Antarctica

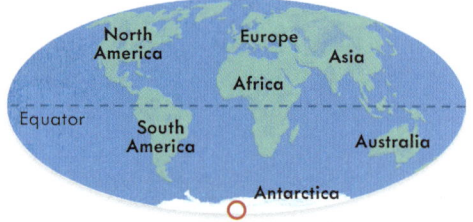

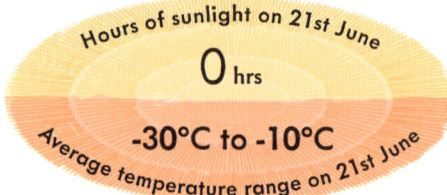

Hours of sunlight on 21st June: **0 hrs**

Average temperature range on 21st June: -30°C to -10°C

LEARN YUI'S LANGUAGE
Japanese

Konnichiwa *(kon-nee-chee-wah)* – hello
Okaasan *(oh-kaa-ah-san)* – mum
Otousan *(oh-toh-oo-san)* – dad
Taiyou *(tai-yoo)* – Sun
Fuyu *(foo-yuh)* – winter

Illustrated by Asako Masunouchi

Konnichiwa, I'm Yui!

Usually I live in Osaka, Japan, but this year we are spending winter at Amundsen-Scott South Pole Research Station. We just call it Pole.

Today is the **fuyu** solstice. **Fuyu** solstice is very different here than it is in Japan. In Osaka it happens in December and the day is very short, but here it happens in June and the **Taiyou** doesn't rise at all. At Pole the **Taiyou** only rises and sets once a year! It's night for six months and then day for six months. The same thing happens at the North Pole, only at the opposite time of year.

What on Earth Books is an imprint of What on Earth Publishing

The Black Barn, Wickhurst Farm, Leigh, Tonbridge, Kent, UK, TN11 8PS

30 Ridge Road Unit B, Greenbelt, Maryland, 20770, United States

First published in the United Kingdom in 2024

Text copyright © 2024 Jen Breach

Illustrations copyright © 2024 What on Earth Publishing Ltd

All rights reserved. No part of this publication may be reproduced or transmitted in any form or by any means, electronic or mechanical, including photocopying, recording, or any information storage or retrieval system, without permission in writing from the publishers. Requests for permission to make copies of any part of this work should be directed to info@whatonearthbooks.com.

Written by Jen Breach

Illustrated by Asako Masunouchi, Cristina Merchán, Daniel Gray-Barnett, Gabi Salem, Gordy Wright, Jannicke Hansen, Mavisu Demirağ, Mus Kabwe, Nabila Adani, Qu Lan, Sakina Saïdi, Tinuke Fagborun, Ubahang Nembang and Vivian Mineker.

Cover illustration by Nabila Adani

Designed by Nell Wood

Jen Breach has asserted their right to be identified as author of this work and Asako Masunouchi, Cristina Merchán, Daniel Gray-Barnett, Gabi Salem, Gordy Wright, Jannicke Hansen, Mavisu Demirağ, Musonda Kabwe, Nabila Adani, Qu Lan, Sakina Saïdi, Tinuke Fagborun, Ubahang Nembang and Vivian Mineker have asserted their rights to be identified as illustrators under the Copyright, Designs and Patents Act 1988.

Staff for this book: Nancy Feresten, Publisher; Katy Lennon, Senior Editor; Andy Forshaw, Art Director; Nell Wood, Designer; Lauren Fulbright, Production Manager.

With special thanks to Joseph Lambert and Patrick Skipworth.

Fact-checking by Satu Fox

All translations and pronunciations have been reviewed by the following native speakers: Ahmed Khedr – Arabic (Morocco); Aysun Demir – Turkish; Elsa Gudiño Carlos – Quechua; Florencia Etchart – Spanish (Argentina); H. Mohammed (Speak2Africa) – Hausa; Mariell Myran – Norwegian; Pradeep Neupane – Nepali; Rui Bi – Mandarin (Chinese Simplified); T. Tlhako (Speak2Africa) – Xhosa; Yuko Okamoto – Japanese; Yusliani Zendrato – Bahasa Indonesian; and Brandi Ramus – Chinuk Wawa. Special thanks to the Chinook Indian Nation and Wolfestone UK.

A CIP catalogue record for this book is available from the British Library

ISBN: 9781913750770

RP/Haryana, India/10/2023

Printed in India

10 9 8 7 6 5 4 3 2 1

whatonearthbooks.com

SOLSTICE

Around the World on the Longest, Shortest Day

Written by
Jen Breach

Illustrated by

Asako Masunouchi
Cristina Merchán
Daniel Gray-Barnett
Gabi Salem
Gordy Wright

Jannicke Hansen
Mavisu Demirağ
Musonda Kabwe
Nabila Adani
QU Lan

Sakina Saïdi
Tinuke Fagborun
Ubahang Nembang
Vivian Mineker

Introduction

I was born and raised in the Southern Hemisphere (the southern half of Earth) and now live in the Northern Hemisphere. In the north and south of the planet the seasons are reversed, so when it's summer in America (where I live now), it's winter in Australia (where I grew up). But there are two special days when these differences are most extreme – the summer and winter solstices.

The solstice is when the Sun is furthest north or south from Earth's equator (the imaginary line that runs around its middle). Earth travels around the Sun on a path called an orbit, and it is also slightly tilted. On the 20th or 21st of June each year, the North Pole is tilted towards the Sun, so it gets more hours of sunlight and longer days. This is summer in the Northern Hemisphere. At the same time, the South Pole tilts away from the Sun, so it is winter in the Southern Hemisphere, and days are shorter and temperatures colder.

Exactly six months later, on the 21st or 22nd of December, there is another solstice, when the South Pole points towards the Sun and has its summer while the North Pole points away from the Sun and has its winter.

Each place in the world experiences the solstices differently, depending on how near or far they are from one or another of the poles. In this book I tell the stories of 14 imaginary children who each live in a different country and experience the June solstice in a different way. We start our journey at the South Pole, where the Sun doesn't rise at all on that day, and head north, getting a little more, a little more, and then a lot more hours of sunlight. We travel through the equator and all the way up to Norway, where the Sun doesn't set on the solstice.

Each story has been illustrated by an artist who has a connection with that place. You can learn about them all on pages 60–61. Let's travel around the world in just one day. Come imagine it with us.

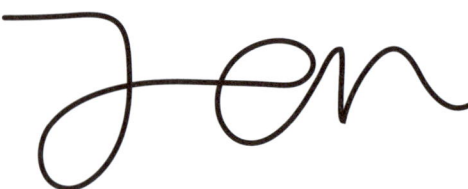

My **otousan** is a botanist – a scientist who studies plants. He looks after the greenhouse that grows all our fresh food. You can't farm outside here as the soil is under about 2 kilometres of ice and the air is so cold that everything freezes instantly. Also, there's no sunshine.

Otousan's plants grow under special lights. He uses a system of water and nutrients called hydroponics to feed them. Because there are no seasons inside, the plants are always growing. Most things at Pole smell like people and machines, so it's nice to go in and smell the plants. Outside smells like nothing. Not that I've been outside since **fuyu** began.

Otousan grows tasty lettuce, herbs, fruit and vegetables – we call them 'freshies'. Cook turns them into our meals along with dried, tinned or frozen foods. Part of **Otousan**'s job is to pollinate the plants so that they grow fruit, since there are no bees, bats or birds here to do it. I'm good at helping with that job. He must also save seeds so next year's gardener will be able to plant a new crop. I help with that, too.

My **okaasan** is an astrophysicist – a scientist who studies how the universe works. There is almost no moisture in the air at Pole and in **fuyu** the darkness is total. So my **okaasan** can do research here that can't be done anywhere else in the world. She listens to the sounds of the Big Bang. That's the explosion that created the universe. We can still pick up leftover signals from it now. They sound like 'shhhhhhhhh'. It's very cool.

Pole is an American-run station, but people come here from all over – scientists from China, New Zealand, Scotland and Chile, engineers from France, Egypt, India and the United States, and Cook is from Canada. There are about 35 people who are spending winter at Pole this year. They all have stories to tell about their homes and their people. I ask what the June solstice is like where they are from. Everyone's stories are different.

Fuyu is very hard at Pole. We are completely isolated. For six months we have to make do with what we brought with us and what is already here. No one will arrive or leave until September, when the **Taiyou** rises and the summer scientists come back with fresh supplies. We don't even have the internet. I had to download all the books and songs I wanted to bring.

The best thing about Pole is the Aurora Australis, the Southern Lights. It's an atmospheric phenomenon that looks like waving curtains of light in the sky – green, purple, bluish, red and pinkish. You can only see them in clear weather, between March and September. They are caused by high-energy particles from the **Taiyou** skidding down Earth's outer atmosphere, drawn to the poles by Earth's magnetic field. The particles gather and create shimmering bursts of light. This reminds me that even though I can't see it now, the **Taiyou** is still there.

Chapter Two

Fire and Light in Tierra del Fuego

Ushuaia, Tierra del Fuego, Argentina

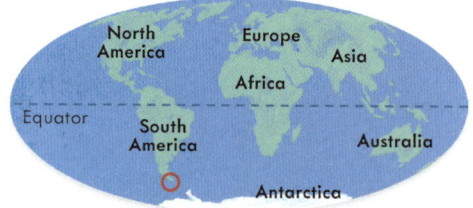

LEARN MATÍAS'S LANGUAGE
Spanish

Hola *(OH-lah)* – hi

Mamá *(mah-MAH)* – mum

Papá *(pap-PAH)* – dad

Sol *(SOL)* – Sun

Invierno *(een-BYEHR-noh)* – winter

Illustrated by Gabi Salem

¡Hola, I'm Matías!

I live in Ushuaia, the world's southernmost city. We're snuggled in between the towering Andes mountains and the exact spot where the South Atlantic Ocean meets the South Pacific Ocean. Lucky me, I can see it from my window.

The **Sol** didn't rise until 10:00am today. It can be so hard to get out of bed on **invierno** mornings. The sunshine is weak and the **Sol** just skims above the horizon. It hardly even feels warm. The shadows stay long and it's night-time again before you know it. I much prefer the summer when it never really gets dark.

My **mamá** is a ski patroller at a busy ski resort. She is away a lot in **invierno** and I miss her. When she's home on her days off, she always has great stories about daring mountainside rescues. She's teaching me to ski, too.

My **papá** is a park ranger at the Parque Nacional Tierra del Fuego, the Argentinian part of a vast wilderness across the bottom of South America. Right now, the park is closed because there has been a lot of heavy snow. The only visitors are sea lions, penguins and millions of birds. They'll live there all **invierno**. Sometimes you can hear the sea lions barking from here.

Mamá is home today though – it's the **invierno** solstice! We call it, 'The Longest Night' and it's a national holiday. The party started yesterday morning and won't finish until tomorrow night. We are so full of good food and warm feelings of friendship that we don't mind the cold. Tonight we won't sleep at all.

Yesterday our friends and neighbours gathered for a meal and we filled our bellies with fire-grilled beef, pork ribs, chicken and potato salad. Then we enjoyed cakes and chocolates. My favourites are the ones with caramel.

Around a fire, we held an important ritual called the 'Burning of Obstacles and Impediments'. On a slip of paper, we each wrote down the things that stopped us from reaching our goals this year. Then we threw the paper into the fire. As the paper burned, we felt freer – we had left our problems behind.

Today, the 21st of June, is filled with more food, more friends and more cheer. Buildings and streets glow with yellow-orange light from flaming torches and big bonfires. Ice skaters glide across the rink where my hockey team practices, under hundreds of sparkling string lights. It's different to the white light of daytime, when the **Sol** bounces off the ice.

Music is playing on every corner – tango, folk and jazz. The air is rich with the smells of pine and bonfire, and with hot street food, too. We snack on savoury pastries called empanadas, gooey ham and cheese tortillas and crisp toasted sandwiches.

At dusk, we head downtown. The whole city is here, along with thousands of tourists. We raise our torches and join the joyful half-walking, half-dancing March of Light.

The Beagle Channel, the passage of water that separates the mainland from the nearby islands, is frozen over and our firelight reflects in the smooth ice. And then, just when you thought there couldn't possibly be more light in our tiny city on the darkest night of the year, the fireworks begin. They scatter whizzing, fizzing pops of colour throughout the sky, lighting up every icy surface. After the last one has popped and fizzled out, we clap and cheer, stomp our cold feet and get back to the music and food.

Chapter Three

Festivals and Art in Hobart

Hobart, Tasmania, Australia

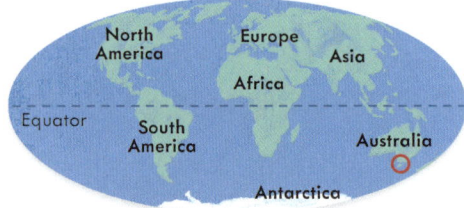

LEARN RUBY'S LANGUAGE
Australian English

G'day (gah-DAY) – hi

Netball (NET-bawl) – a ball sport played on a court by two teams of seven

Milo (MIGH-low) – chocolate-flavoured malted milk powder that is usually mixed with milk or hot water

Ogoh-ogoh (O-goh O-goh) – Indonesian word for a statue, usually a demon, built for a parade called Ngrupuk (NG-ru-puk)

Illustrated by Daniel Gray-Barnett

G'day, I'm Ruby!

It's been cold and rainy in Hobart today. Even though the Sun rose at around 7:45am, it was hidden by grey clouds all day. I woke up in the dark and it was dark again by the time I got home from school and **netball** practice at 4:45pm. That's winter in Tasmania!

But today is not just any old winter day. It's the 21st of June! Our city celebrates midwinter with a huge public arts festival that lasts for weeks. All over Hobart, the buildings are lit up with lights of every colour.

There are bonfires, brilliant works of art that are made out of lights and laser shows. The whole town turns up for the festival, and tons of tourists, too.

Despite the cold and rain, there is music and food everywhere. There is a hum of excitement in the chilly air. In public spaces, people huddle around braziers and gas heaters, before taking in free artwork and concerts.

My mum works at the contemporary art museum that puts on the festival. She has to help figure out how to set up art installations that need thousands of lightbulbs or candles or that need to be safely set on fire. My stepdad is a chef in a cute cafe most of the year (he makes great avocado toast), and during the festival he runs the cafe's stall at the Winter Feast. Mum and I are on our way there now for a tasty snack that will warm us up from the inside.

There are so many sounds and smells — jazz and pizza, indie rock and hog roast, folk music and burgers, punk and charred veggies.

Mum and I stop to buy some fresh jam doughnuts. The sweet jam slides down our throats all the way to our toes. Hopefully we can also find a steaming mug of **Milo** to warm us up.

We learned about the winter solstice in school. We learned that in ancient times people all around the world would pray for the Sun to come back again, because it would bring good harvests and protect against evil spirits. In Japan, people celebrate by lighting fires and taking baths scented with a citrus fruit called yuzu. The indigenous Zuni peoples in New Mexico, U.S., do a traditional dance called Shalako that lasts for five days.

One of my favourite traditions of our festival is coming down the main street now. It's the **ogoh-ogoh** parade. Local people of Indonesian descent build and paint a big wooden statue and march it through the festival, waving flares and banging drums and pots. At the end of the march, they set fire to the statue and dance, cheer and sing while it burns. Usually, Indonesian Hindus burn an **ogoh-ogoh** in March, on the first day of their new year, to spiritually heal the environment. Here they do it on the 21st of June as an art piece. It is really fun! Mum and I dance and cheer along and look forward to tomorrow, which will be a longer day.

Chapter Four

Whales and Sardines in Cape Town

Cape Town, South Africa

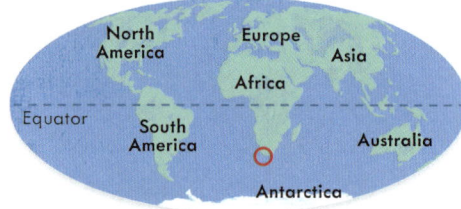

LEARN NOBOMI'S LANGUAGE
Xhosa (klo-sah)

Molo *(MAUL-low)* – hello
Umama *(oo-MUH-MUH)* – mum
Utata *(oo-TUH-TUH)* – dad
Ilanga *(ee-LAA-ngah)* – Sun
Usisi *(oo-SEE-SEE)* – sister
Umakhulu *(oo-MA-koo-loo)* – grandma

Illustrated by Musonda Kabwe

Molo, I'm Nobomi!

This will be our second winter in Cape Town. We used to live in Port St Johns, a little town on the wild and rugged coast of Eastern Cape. Cape Town is a big city on the calm and sunny Western Cape.

I miss two things about Port St Johns: my friends and the sardine migration. At this time of year a group of sardines swims hundreds of kilometres along the east coast to their spawning grounds, where they lay their eggs. Millions of sardines travel together – the group can be six kilometres long!

The sardines shimmer and shine under the **ilanga**. Thousands of animals gather to eat them: bigger fish, seals, sharks, dolphins and more! My **utata** used to run diving tours for tourists, and the sardine run in June was his busiest time. But he always had time to take me and my **usisi** out to see them.

Now we live in Cape Town and **Utata** guides tours for animals a bit bigger than sardines: whales! Today is the solstice, but it's the same busy day as much of the rest of the winter – June is the peak time that whales migrate along South Africa's west coast between their feeding and breeding grounds. When I'm on a tiny boat in the ocean and a whale breaches the water, I feel the way a sardine must have felt when it saw me – a little scared but totally amazed! This year I am doing my SCUBA diving certification, so hopefully I'll get to swim with whales one day. Another thing I would have missed about Port St Johns was my **umakhulu**, but she came with us to Cape Town!

There are two things I like better about Cape Town than Port St Johns. First, **umama** isn't away as much. She's an ornithologist, which means she studies birds. She used to be away for months at a time. She would stay in tents in national parks, spending all day looking up into the trees and counting birds.

Now she's a professor of ornithology at a university in Cape Town. But whenever we're outside she's still looking up into trees and counting birds. **Umama** and **Utata** tell goofy jokes about how he's a fish and she's a bird, but somehow they managed to fall in love.

Umama's new job is why we moved. It was really hard to say goodbye to my friends and the sardines in Port St Johns. Cape Town is so big, I was scared it might swallow me up like a whale swallowing a sardine. But after my first day at school, I knew it would be okay.

The second thing I like better in Cape Town is school. Here it doesn't matter that I am biracial. I'm not even the only biracial kid in my class! No one really talked about it in Port St Johns, but in a way that felt really different to how no one talks about it in Cape Town.

Marriage between people of different races became legal more than 35 years ago, but it still doesn't happen all that often. My other grandparents don't talk to my parents anymore, because they didn't agree with them getting married. It makes me sad, but also so happy to be smooshed up in the hugs of my **umakhulu** who lives with us.

Umakhulu makes us lots of tasty food — I especially love her sweet apricot jam pudding and milk tart. The food she makes is full of love, even her Bunny Chow, which is a hollowed-out loaf of bread with curry inside. All the kids at school want to trade their snacks for mine!

Chapter Five

Incan Festivities in Ingapirca

Ingapirca, Ecuador

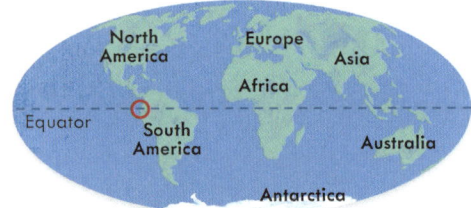

Hours of sunlight on 21st June
11hrs 55min
23°C to 29°C
Average temperature range on 21st June

LEARN TUTA'S LANGUAGE
Quechua (KEE-chu-wa)

Imanalla *(ima-NA-lia)* – hello

Mama *(ma-ma)* – mother

Tata *(ta-ta)* – father

Inti *(in-ti)* – Sun, also the name of the Incan sun god

Inti Raymi *(in-ti RAY-mee)* – Sun Festival

Illustrated by Cristina Merchán

Imanalla, I'm Tuta!

I live with my **tata** in the village of Ingapirca, 270 kilometres south of the equator. My **mama** lives in Otavalo, a city that's just north of the equator. For her, today is the summer solstice, but for me it's the winter solstice! We're only a few hours away from each other, but because we are on different sides of the equator, we are in different seasons! The winter solstice in Ingapirca is not that different from the summer solstice. During the summer solstice in December, the **Inti** will rise only about 20 minutes earlier than it did today.

The June solstice is the last day of the Incan year in Ecuador. It is our most important festival. We celebrate everything **Tata Inti** has given us, and ask to be blessed in the coming year.

It all started this morning right after midnight. My **tata** and I went to the waterfall with others from our community to cleanse our spirits. We reconnected with the goddess of the Earth, who is called Pachamama. Pachamama takes care of us, so the ritual is important. It was very early and the water was very cold!

Afterwards we dressed in our mostly brightly coloured traditional clothes – crisp white blouses and shirts, shawls and ponchos, many-layered skirts and shaggy chaps over our trousers. We walked from our little village up the hill to the grand ruins of Ingapirca.
That's where we wait for the **Inti** to rise.

The population of our town more than doubles for **Inti Raymi**. People who have moved away come home to celebrate with their families, and Quechua people come from nearby towns to be in this special place. There are some tourists, too, but not many.

Ingapirca sits 3,200 metres above sea level and is pretty remote and hard to get to. Most tourists prefer the festivities in Otavalo or Cuzco, Peru. Cuzco was the capital of the Incan Empire, and I've seen videos of the **Inti Raymi** pageant there. Dozens of actors in Incan costumes re-enact the ritual as it might have been in the past. The actor playing Sapa Inca, the emperor of the Incan Empire, is carried on a golden chariot to where the Sun Temple used to be. He makes speeches about prosperity, good harvest and our gratitude, and pretends to sacrifice a llama!

In Ingapirca the Sun Temple was built by my ancestors. When Spanish colonists invaded Ecuador in 1531, they knocked down the temple and banned our festivals and rituals. Some Incan people kept practicing **Inti Raymi** even though they could have been killed if they were caught. But still, they taught their kids about **Inti Raymi**. Those kids grew up and taught their kids, who taught theirs. My Dad taught me and I'll teach my kids, too. We remember all of this while we wait for the **Inti** to rise on the solstice.

The rising **Inti** is our cue to begin celebrating. We honour **Tata Inti**, Pachamama and each other by sharing a meal – potatoes, beans, goose and corn cooked in lots of different ways. Then we dance and sing and eat some more. The air is filled with flutes, guitars, drums, pounding feet, cheers and song all day and all night.

Chapter Six

River Games in Pontianak

Pontianak, West Kalimantan, Borneo, Indonesia

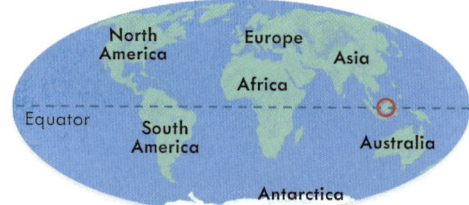

Hours of sunlight on 21st June
12 hrs
23°C to 29°C
Average temperature range on 21st June

LEARN DIMAS'S LANGUAGE
Bahasa Indonesian

Halo *(haa-lo)* – hello
Ibu *(ee-boo)* – mum
Ayah *(eye-ah)* – dad
Matahari *(maa-tuh-haa-ree)* – Sun
Adik *(ah-dik)* – younger brother
Qur'an *(kaw-raan)* – the central text in the Islamic religion

Illustrated by Nabila Adani

Halo, I'm Dimas!

I live in the capital city of Pontianak, right on the equator. You know how the Earth is tilted on its axis so that in some places the summer and winter solstices are a big deal? Well, that tilt doesn't mean anything at the equator.

Because the equator never tilts towards or away from the **Matahari**, here every day is the same length, even when it's a solstice. The **Matahari** only takes a few minutes to rise in the morning. And exactly 12 hours later it only takes a few minutes to set. It's the same every day, all year round.

In other parts of Indonesia, like the coasts and mountains, they have a rainy season and a dry season. In Pontianak we only have one season and one type of weather: hot and humid with afternoon rain. Sometimes, puddles don't evaporate because there is already so much moisture in the air. It always smells like rain, freshness and green.

My family runs a rice farm just outside the city, and in this weather it seems as if the rice grows itself. Everything grows itself here. And grows big. Including me!

Flowers, grasses and trees are everywhere and are home to birds and insects that squawk and buzz and drone all day and all night. We're all used to the noise though.

I've seen videos of snow where people wear huge boots, woolly hats and puffy parkas. Their noses glow red in the frosty air. I've only ever worn shorts and short-sleeved shirts or tees. Except once, when my auntie got married. I was the ring-bearer and had to wear a suit to the wedding. It was scratchy and awful!

If it's not too rainy after school and I don't have basketball practice or **Qur'an** class, I'll head to the riverside with my friends. I love jumping in – the splash and whoosh of breaking through the water makes me feel like I'm flying. We often row boats up and down the river. I like to surprise my friends and tip their boats over. We all end up drenched. Today the river is calm, so I float on my back, basking in the **Matahari** as it takes its unchanging path across the sky. My ears are underwater and filled with funny gurgling sounds. It's quiet without the chirp and buzz of bugs – peaceful, but also a little eerie.

My **ayah** is the manager of a hotel downtown and my **ibu** is a pastry chef in the hotel restaurant. They work very hard. Our babysitter is called Indah and she is home all day looking after my **adik**, who is too young for school. Indah works hard, too. But she always has time to play hide and seek or video games if we help her fold the laundry. After I beat Indah (I always win), we get started on dinner. In the mortar and pestle, I pound chillis, lime and sugar into a spicy sauce. My brother peels eggs for the delicious gado-gado salad while Indah chops fresh vegetables. Peanuts go in, too.

In a place where every day is the same, Saturday is definitely my favourite. Saturday is the day before our rest day. Indah lets us stay up until **Ayah** and **Ibu** come home from the hotel. **Ibu** will bring us her famous durian fruit tarts for a late-night treat. The creamy, grassy sweetness dissolves in my mouth and I think she must be the best baker in the city.

Chapter Seven

Sips, Snacks and Shade in Barage

Barage, Nigeria

Hours of sunlight on 21st June
12 hrs 55 min

23°C to 32°C
Average temperature range on 21st June

LEARN SANI'S LANGUAGE
Hausa

Sannu *(sah-NOO)* – hello
Iyali *(ee-yah-LEE)* – family
'Yan uwa *(eeyun OOWAH)* – cousins
Rana *(rah-NAH)* – Sun
Kakanni *(ka-ka-NEE)* – grandparents

Illustrated by Tinuke Fagborun

Sannu, I'm Sani!

I live in Kano, a big city with paved streets, electricity and the burble of English everywhere. We speak Hausa to each other, but English is the language of business and tourism, and there are plenty of both in Kano.

Today, my **iyali** and I are driving to the small village of Barage. On the way we collect our friends and **'yan uwa**. It's a three-hour drive on a twisty road that runs through the dry land. Barage is a world away from our apartment in the city. No one speaks English there.

Barage is the hot, dusty town where most of my **iyali** and I were born. It has always been a small village of goat herders. My **kakanni** will never move away. We keep trying to get them to move to the city with us, but they refuse to even visit! They always have **'yan uwa** around to help out. My great-grandparents are both turning 70 this weekend, which is why our van is fuller than ever. It's quite an occasion!

This weekend is also the summer solstice, but we don't really call this time summer. We are right between the dry and wet seasons. It is not as hot as it was last month and not as muggy as it will be next month. I know that some places have hot and cold seasons, but in Nigeria it's always hot.

We learn about the solstice in class – how the tilt of the Earth makes seasons. The start of each season can be a solstice, when it is the longest or shortest day of the year. Or it can be an equinox, where the hours of daylight and night are equal. It's cool to think about places where the **Rana** doesn't set on the summer solstice, or places that have a cold winter solstice with snow and bonfires. In Nigeria, the solstice is just like any other day. Except for us it will be a birthday party!

In Barage, power comes from solar panels, which charge generator batteries. There is plenty of sunlight all year round, but batteries are expensive and most people only have one or two. We can have power for cooking and tools for part of the day and that's it. And, there's no air conditioning in Barage. Instead, people tend to stay in the shade. One of my teachers has travelled the world. She says the **Rana** doesn't shine anywhere in the world the way it does in Nigeria. Here, the **Rana** is big and fiery orange and bakes everything it touches.

When we arrive in Barage, we all tumble out of the van and into the enveloping embraces and welcoming kisses of our extended family. Grandpa and Great-grandpa wave from the shade of the biggest tree in the village, where they sit and gossip with old and new friends.

Grandma and lots of aunties lay out snacks in the shade. We munch crunchy, fried peanut balls (Great-grandpa's secret recipe), nuts and cucumbers, roasted maize and coconut. And we sip on refreshing hibiscus tea. We talk and laugh and greet more and more vans full of **iyali** as they arrive and the **Rana** sets.

Chapter Eight

Momos and Chiya in Pokhara

Pokhara, Nepal

Hours of sunlight on 21st June
14 hrs
23°C to 29°C
Average temperature range on 21st June

LEARN LAKSMI'S LANGUAGE
Nepali

Namaste (na-maste) – hi
Āmā (ah-ma) – mum
Bubā (bu-ba) – dad
Sūrya (sur-yah) – Sun
Momos (mo-mos) – dumplings filled with vegetables or meat

Illustrated by Ubahang Nembang

Namaste, I'm Laksmi!

When the **Sūrya** rose this morning it made the Annapurna mountain glow orange-pink. My **āmā** and I were already at work in her store at one of the best viewing spots above Pokhara. If the tourists are up before the **Sūrya**, so are we. But we don't have to travel far to get to work – we live in the room right behind the kiosk! **Āmā**'s shop is open every day from before dawn until night. So today, the 21st of June, the wet season solstice, is the longest working day of the year.

Even though we have to get up earlier, the wet season is my favourite. It usually rains all night, but mornings are clear and everything sparkles. There is no green like wet-season green in Nepal. All the tourists say so. The air smells kind of green, too. At least until **Āmā** puts the chiya spices on to steep. Chiya is a spiced tea and it makes the air smell like **Āmā**'s special blend of cardamom, ginger, black pepper and fennel. We both work up the dough and filling for the morning's **momos**.

I don't know anyone quite like my **āmā**. My **bubā** was a Gurka (a soldier) in the Nepali army. He passed away when I was a baby. I don't remember him, but we have a photo of him in uniform. He was very handsome. When he died, everyone expected **Āmā** to do the usual thing that widows in Nepal do: disappear from view and live in poverty. But that didn't suit **Āmā**. With me strapped to her back, she rebuilt a run-down shack into this kiosk. The first things she offered for sale were chiya and **momos**, but now she has all kinds of western snacks, too.

Like many Nepali women, **Āmā** can't read or write. She's smart, though – smart enough to run a business and to pick up English, French, German and Spanish from the tourists. She's really good at maths, too. She had taught me some numbers before I even started school.

School finishes in the middle of the day and as soon as it does, I rush home. Tourists flop under our awning for relief from the high, hot **Sūrya**. There's always a buzz of people at the kiosk sipping soft drinks or chiya, eating snacks or **momos** dipped in spicy tomato chutney.

I quickly change out of my school uniform and hang it up ready for tomorrow. I throw on light clothes (much cooler than my itchy uniform!) and bustle around collecting cups and bowls to wash. I put some more spices on to steep and prepare another batch of **momos**, and before I know it the **Sūrya** is setting and more tourists arrive.

It's a very dramatic view tonight. Dark grey storm clouds roll over the lake and hills, with radiant orange-yellow sunbeams behind them. The air smells like rain. But the rain holds off until the sky is dark and the tourists have left.

That's when we prepare our own evening meal of rice and lentils full of warm spice. While we cook and eat, we tell each other everything we learned today. I try out some new English words and phrases, which **Āmā** repeats. It's late by the time we turn off the propane light and go to sleep. We know that we will soon be up again and the days will start getting shorter.

Chapter Nine

Work, School and Home in Marrakesh

Marrakesh, Morocco

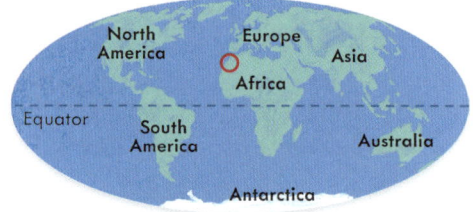

LEARN AHMED'S LANGUAGE
Arabic

Marḥaban *(mar-ha-ban)* – hi

'Um *(Umm)* – mother

'Ab *(UH-boo)* – father

Shams *(shams)* – Sun

Sbah lkhir *(sa-BAH al-khair)* – a respectful way of saying 'good morning'

Illustrated by Sakina Saïdi

Marḥaban, I'm Ahmed!

Welcome to the warrens of Souk Semmarine, the most famous market in all of Marrakesh! It is only 8:00am so the market isn't open yet, but it's already bustling with vendors setting up their stalls. The market will only get busier when the gates open! The air is already thick with the smell of spices, oils and leather. It smells like home.

Every morning, I help my **'Ab** and **'Um** set up their ceramics stall. We arrange tagines (clay cooking pots with cone-shaped lids), bowls and plates on rugs on the pavement. We say **'sbah lkhir!'** to the other vendors as they arrive. Setting up the shop in summer is sweaty work. Especially today, on the solstice. The **Shams** has been up for hours and it's already hot.

We set up **'Ab**'s heavy pottery wheel and **'Um**'s glazes and paints. **'Um** only paints some of **'Ab**'s pieces – the tourists snap those ones up. But most of their pottery is unpainted and bought by locals to cook with. **'Um**'s colours are stunning – she paints layers of floral and geometric patterns she knows from the traditions of her ancestors, or that she invents on the spot.

Once the shop is set up, I race through the city and make it to school just in time. I enjoy learning, so the day flies by. After school and football practice, I head straight home to the roof of our house to feed our pigeons. They coo when they see me and I coo back. I make sure the birds are cool and have enough fresh water. I am training two pigeons to be racers.

After a while, my younger sister comes to ask me to make her a snack. Once she is fed, I put supper on to stew, in one of '**Ab**'s tagines, of course! Cubes of beef, aubergine, tomatoes, onions and a handful of spices. It smells like the market. Then it's time for homework. I have to make sure my sister does her homework, too.

I am working on English language vocabulary exercises. English is really hard. It has so many weird rules! Like most people in Marrakesh, I already speak a bunch of languages.

la glace

At home we speak Tamazight, the language of the Amazigh, Morocco's native people. **'Ab** is Amazigh, like 40 per cent of Moroccans, and **'Um** is Arab-Amazigh, like most of the rest of us. But movies, TV shows, books and school lessons are in Arabic or French. With friends and neighbours we speak Darija, a kind of Arabic only spoken in Morocco, or Tamazight or sometimes Spanish. It's not uncommon for me to speak seven languages in a single day.

'Ab and **'Um** are tired when they get home at 9:00pm, after the market closes. But they are never too tired for the supper I have made. Afterwards, we take a stroll for coffee and cake or ice cream.

Today, on the summer solstice, we enjoy the long twilight as we lick our sticky fingers. We linger in the last light and wish our neighbours who are doing the same 'good health'.

Chapter Ten

Busy Days in Beijing

Beijing, China

Hours of sunlight on 21st June
15 hrs
26°C to 32°C
Average temperature range on 21st June

LEARN ZHANG JING'S LANGUAGE
Mandarin

Nĭ hăo *(nee-how)* – hi
Mŭ qīn *(mu-chin)* – mother
Fù qīn *(fu-chin)* – father
Tài yang *(tie-yang)* – Sun
Bāo zi *(bow-ze)* – steamed buns that can be be filled with a variety of ingredients, such as ground meat, egg, vegetables and herbs

Illustrated by QU Lan

Nĭ hăo, I'm Zhang Jing!

It's another hot, humid, smoggy day in Beijing! I'm fast asleep when the **Tài yang** rises. I wake to the sound of my alarm clock and look out of the window to see an orangey-brown-grey haze. Even though our apartment is high up, I can't see far across our neighbourhood because of the smog. I quickly get dressed for school, which is easy as I wear the same uniform every day.

My **mǔ qīn**, **fù qīn** and I eat breakfast together and talk about the day ahead. We have pillowy **bāo zi** stuffed with egg and chives. My parents ask me to work harder at school and pay attention. They often ask me this. Before I know it we are running late. We grab another **bāo zi** each and dash out the door. I hurry to my bus stop and they hurry to the underground train.

My **fù qīn** works in a big office building downtown – boring! My **mǔ qīn** is a panda nanny at the Panda House in Beijing Zoo. Pandas need to be played with a lot so they don't get sad, just like kids. They also love to nap. I have been to her work twice. I stripped bamboo for the pandas to eat. I even got to pet one. Panda fur doesn't feel as soft as it looks. It's thick and a bit scratchy. I want to be a panda keeper when I grow up. I love them and the way the black fur on their ears and eyes looks!

Mǔ qīn and **Fù qīn** want me to have an office job in the city. My grades aren't as high as they would like. They want me to take up a sport or an instrument so I have a better chance of getting into high schools. But I don't mind if I don't go to high school. I just want to play with pandas.

School classes start at 8:30am and end at 5:00pm. I have tutoring after that, but I'd rather be playing video games with my friends. Today is the summer solstice, which means it will still be light when I get home.

But the solstice means something else, too: soon it will be the Dragon Boat Festival! It's my favourite holiday of the year. It's all I can think about in class and all I can talk about at lunch. I love everything about the Dragon Boat Festival.

We get three days off and we clean our flat from top to bottom. We hang sprigs of a plant called mugwort around the flat to keep away disease for another year. **Mǔ qīn** will give me a silk pouch that is filled with herbs to wear around my neck. It keeps evil spirits away.

My **fù qīn** and his dragon boat team compete in the race, and we always go to cheer for them. They never win, but we will celebrate anyway and eat sticky rice triangles with tasty savoury fillings. How can I think about school when there's a festival coming up?!

Chapter Eleven

Cats and Mosques in Istanbul

Istanbul, Turkey

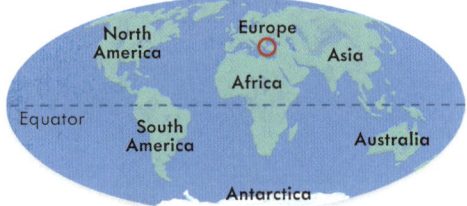

LEARN ZEYNEP'S LANGUAGE
Turkish

Merhaba *(mer-ha-BAH)* – hello
Güneş *(gue-NESH)* – Sun
Anne *(ahn-NAY)* – mother
Baba *(baa-BAH)* – father
Çaydanlık *(chai-DHAN-laehk)* – a double stacked teapot used to make Turkish tea

Illustrated by Mavisu Demirağ

Merhaba, I'm Zeynep!

I live in Istanbul and today got up with the **Güneş** at 5:30am. I'll stay up with it all day, which will be a long one, since it's the summer solstice. The **Güneş** won't be gone until 8:45pm. It's also the last day of school before the summer holidays. I'm so excited! I start the **çaydanlık** and grate tomatoes for the breakfast stew. I put out food and water for the stray cats, and a cardboard box for them to play in. I wave **merhaba** to my neighbour.

By then, my **baba** is up. He works long hours in the summer. He is a stonemason, like his **baba** before him and like his brother, too. Stonemasons build structures out of stone. If I had a brother, he probably would have become a stonemason. Not me. I'm going to be an architect. After breakfast, **Baba** packs lunch, including a little extra for the Hagia Sofia cats. His favourite is a tortoiseshell that he named Zeynep after me.

The Hagia Sofia is an impressive building that the people of Istanbul are very proud of. It's where my **baba** works. It took 11,000 stonemasons to build it in the year 537, and now it takes dozens more to keep it in good repair. It gleams pink, orange, sandy or almost-white, depending on the light. **Baba** and the other stonemasons are like ants on the outside of the building, constantly running here and there, repairing and maintaining the beautiful stone. I love his rough hands, brown from the stone and long hours in the **Güneş**.

It's not really allowed, but once, he took me up onto the scaffolding where he was working near the dome. The city's noise melted away into a hum and rumble. My beautiful city shone and sparkled. I saw the towers and domes of grand palaces and mosques, the red terracotta roofs of humble homes and the steel and glass of modern skyscrapers. I could see the whole city's history from the Egyptian obelisk to the Roman Hippodrome and Blue Mosque. I could imagine the buildings I was going to design dotting the cityscape. I could see all the way to Asia...

... OK, Asia actually isn't that far. Istanbul spans two continents. I live in Europe, but every day my **anne** and I take the bus over the long Fifteen Martyrs Bridge, high above the wide Bosphorus strait to school in Asia.

But not after today! For three whole months I will play with **Anne** and my friends every day. We'll go to the beach to swim, play in the park and eat and laugh. And I will not have one single lesson! Maybe **Anne** and I will go see a match in the women's football league. If I wasn't going to be an architect, I would be a professional football player.

Maybe **Anne** and I will visit **Baba** at work and we will pet the cats. We'll bring a picnic of homemade spiced meatballs, rice wrapped in vine leaves and soft, chewy fillo-pastry pie. And, of course, some baklava, which is a dessert made of filo pastry, chopped nuts and lots of honey. Those are **Baba**'s favourite. Mine, too. Maybe **Baba** will even sneak us up onto the scaffolding again to admire the city.

Chapter Twelve

Salmon and Sunshine in Ilwaco

Ilwaco, Washington, United States of America

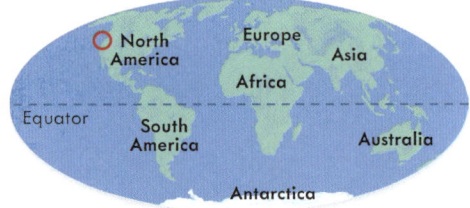

Hours of sunlight on 21st June
15 hrs 45 min

11°C to 15°C
Average temperature range on 21st June

LEARN RILEY'S LANGUAGE
Chinuk Wawa

Łaxayam *(lak-sa-wy-em)* – hello
San *(san)* – Sun
kʰwaɬ *(kwahl)* – aunt
kʰul-iliʔi *(bcole-illa-hee)* – winter
wam-iliʔi *(waum-illa-hee)* – summer

Illustrated by Vivian Mineker

Łaxayam, I'm Riley!

Here in Ilwaco, lots of people work on fishing boats or in fish processing factories. The whole town smells tangy like fish and the ocean.

I live with my **kʰwaɬ**, who is the skipper of an all-female fishing boat crew. Many of her crew live here only for the peak fishing season during the warm months, but my **kʰwaɬ** and I live here year round. Her boat is called April Ray and my **kʰwaɬ** loves to take her out on the water to catch black cod and halibut.

During **kʰuɫ-iliʔi** the crew often goes to work on rich people's boats. They really enjoy this because it means they can chase the **San** to tropical beaches. I have a really good collection of postcards, knick-knacks and snow globes that they have sent me from their travels. One day I'll visit all those places, too. But I'll always come back to Ilwaco when the fish do.

On the **wam-iliʔi** solstice the **San** doesn't completely set, but skims just below the horizon. Sunsets can last a really long time in **wam-iliʔi**. It's only true night-time for about two and a half hours. Most of the night it is not quite light enough to read, but not dark enough to turn on the lights. We won't have a proper night for another couple of months.

June is the busiest time of year for fishing. The salmon come back to the coast from the deep sea. They wait until the conditions are just right to run upriver to lay eggs. While they wait, our port is full of fish. The squawking sea birds are noisy. Birds such as petrels, shearwaters, terns, skimmers, bald eagles and sea eagles are always diving into the clear water for a catch. Albatrosses are the best to see diving – they are so big and powerful that all the other birds get out of the way. Sometimes the people do, too.

When my **kʰwaɬ** and her crew bring the catch into harbour, we move quickly. We sort the fish into the ones that will stay whole and the ones we will fillet. We gut, fillet, vacuum pack and snap freeze them right there, just hours out of the water. King salmon are huge, they can

weigh around 13 kilograms – that's more than a watermelon! I can barely lift one. So I stick to sockeye salmon, which are about 3 kilograms. I'm sure I'll soon be able to sling the king salmon around like the rest of the crew.

My school is on **wam-iliʔi** holiday, but I have plenty to do even though I'm too young to go out on the April Ray. I can help to prep the bait for the next day's haul or work in the garden where we grow lots of food. While the boat is out, I weed and pick fruits and vegetables to eat and preserve. We have beetroot, beans and broccoli, cabbage, corn and kale, spring onions, spinach and summer squash. Sometimes I can talk my friends into helping me. But they have their own work to do. We'll get a break in the **kʰul-iliʔi**.

Chapter Thirteen

Sunrise at Stonehenge

Stonehenge, Salisbury Plain, United Kingdom

Hours of sunlight on 21st June
16 hrs 30 min
10°C to 19°C
Average temperature range on 21st June

LEARN EMILY'S LANGUAGE
British English

Henges *(HEN-jez)* – prehistoric monuments that are usually made of stone or wooden slabs arranged in a circle

Druid *(DROO-id)* – name for or a priest or holy person in the ancient Celtic religion

Pagan *(PAY-gun)* – someone who worships Earth or nature, and a number of different traditional gods

Wiccan *(WICK-un)* – someone who belongs to a Wicca religion, based on the power of nature

Illustrated by Gordy Wright

Hello, I'm Emily!

In summer the weather in Southern England could be just about anything – rainy, grey, bright, muggy, hot, cold! Today, I'm up before the Sun, and I can see the stars, a good sign that the skies will be clear for sunrise. The air is crisp and there is dew on the ground, so I bundle up, put on my wellies and head out with my family. We are walking the kilometre or so from our village to Stonehenge, which is a really old **henge** made of large stone slabs. Today is a special day because it's the summer solstice – the longest day of the year.

Last night was the shortest night of the year. It didn't get totally dark because the Sun stayed just below the horizon, so some of its light could still be seen. However, we still need torches when we leave the house this morning.

Our torch beams catch rabbits' fluffy tails as they dash away down the lane. In the distance, we can see the white headlights and red rear lights of cars on the motorway. The swooshing of the cars is the only sound in the still morning air. That is, apart from all the racket we are making, singing and laughing as we walk.

I love Stonehenge. So do my parents – they met there, got engaged there, and would have got married there if they could. Today is their anniversary. On their wedding day we all saw the Sun rise at Stonehenge, and we've been doing the same every summer solstice since.

My stepmum is an archaeologist, she can talk about prehistoric finds for hours! It's not a family holiday unless she takes us kilometres out of the way to look at some 3,000-year-old stone circle. Dad's obsessed with Stonehenge, too. He's a tour guide here two days a week and loves his job.

This morning, on our way to the Stone Circle, we pass the Heel Stone. For me, there's something special about that stone. Every other stone at Stonehenge – the ones in the outer circle and the inner horseshoe and the bluestones – have been shaped with tools. The Heel Stone, though, is rough and natural. It stands outside the Stone Circle, different and special.

No one knows why this stone has been left rough, because Stonehenge is prehistoric. That means that there are no written records from or about the time it was built. The only hint we have about why Stonehenge was made is Stonehenge itself. And it's not telling.

We stand among the stones, with thousands of others. People sit on slightly damp picnic blankets or stand huddled in groups or by themselves. Some **Druids**, **pagans** and **Wiccans** dress in traditional clothing, and it makes the gathering feel very festive.

Everyone faces the northeastern sky, where the navy blue twilight dissolves into yellow-blue dawn. And then, around 4:45am, we all gasp. Gold sunlight spreads across the field. The long shadow of the Heel Stone appears, touching the centre of the Stone Circle, something that only happens on the summer solstice. It always takes my breath away.

Chapter Fourteen

Up All Night in Longyearbyen

Longyearbyen, Svalbard, Norway

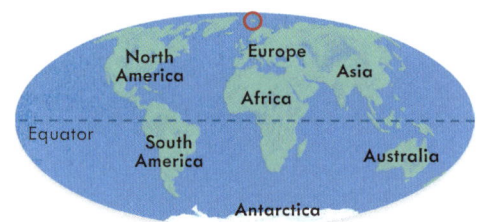

LEARN KJERSTI'S LANGUAGE
Norwegian

Hei *(high)* – hi
Mamma *(ma-ma)* – mum
Mor *(mawh)* – mother
Solen *(SO-ween)* – Sun
Besteforeldre *(BA-sta-fur-aldra)* – grandparents

Illustrated by Jannicke Hansen

Hei, I'm Kjersti!

I live inside the Arctic Circle, in the world's northernmost town: Longyearbyen. Here, the **Solen** doesn't set on the summer solstice. That's today, the 21st of June. Actually, the **Solen** hasn't set since the 19th of April, and won't until the 23rd of August! We are very far north.

Both of my grandfathers worked as coal miners, like almost everyone's grandfathers around here. Both of my grandmothers also worked for the coal mining company, one as a caretaker and one as a cook in the canteen.

The old mine is now the Global Seed Vault, a huge freezer containing seeds from more than a million plants. There's no electricity there – the icy mountain itself keeps it cold. It's the biggest vault of its kind in the world! We once went on a school trip there and saw rows of shelves filled with lots of sealed black boxes. We got to see inside a box that was being put there for storage. Inside were a hundred carefully labelled silver pouches filled with seeds from Burundi. I bet my **besteforeldre** would have been amazed by the vault.

Tourists say it's amazing that the **Solen** doesn't set, but it's normal for me. We use blackout curtains and eye masks to pretend it's dark at night so we can sleep. But it's hard to feel sleepy when we could be playing outside all night. Tourists also say it's amazing that we can sometimes see polar bears from our windows. That feels normal too. There are lots of polar bears on the islands of Svalbard. Since the coal mine closed, they don't mind coming closer to our town.

Today, the night sky looks exactly the same as the day sky. It's any one of dozens of shades of blue. I'm glad it's not grey. It's often cloudy and grey and raining here. It is still cold, even though it's summer.

Because of the rain and sunshine, the fields are bright green and covered in wildflowers – buttery-white Svalbard poppies, star-shaped purple saxifrage and fluffy pink heather. It's so different from the silent darkness of the winter solstice. It smells different, too.

The midsummer festival happens on St John's Eve, on the 23rd of June. Like the rest of our town, my **mamma** and **mor** will have the day off work and I won't have to go to school. Many places in Norway have new traditions for midsummer, but in Svalbard we stick to the old ways. We gather lots of wood for bonfires. Some years we can't have bonfires because they might start brush fires, but this summer has been wet, so it's okay. We might have to dance in our wellies, but we can stack the bonfires high. We enjoy a feast with family and friends, and at 9:00pm we light the fires to ward off evil spirits for another year.

Meet the Illustrators

Chapter One
Asako Masunouchi

I was born in Japan, studied in the UK and now I live in Greece. I have been working as a freelance illustrator since 2005. I like the way that this book shows how people spend their time on the solstice. It took me on an imaginary trip as I created my artworks. My illustrations represented Japan through Yui's character. I thought about what she likes, what she collects and what she took to Antarctica with her. I took inspiration for Yui from my own childhood and from my daughter.

Chapter Two
Gabi Salem

I am an illustrator, graphic designer and university professor. I grew up in Buenos Aires, Argentina, and still live there today with my boyfriend Guille and my little puppy Nina. This project is very special to me because when I was a kid my dad spent some time working in Ushuaia, Tierra del Fuego ('Land of Fire' in Spanish). I received lovely postcards from him showing a beautiful town in a bay full of boats and an amazing autumn landscape, with a rich, red forest. I always wondered what it would be like to visit, and a few years ago I finally did. I loved it! It is a very special place to me. I hope you can get to know Ushuaia someday, too!

Chapter Three
Daniel Gray-Barnett

I'm an illustrator and children's author from the Huon Valley in Tasmania, Australia. I grew up in Sydney but moved to Tasmania several years ago, where I live in the country (about 40 minutes south of Hobart). Life in the country is a lot quieter than busy Sydney – we are more connected to nature and really experience the seasons. In our valley, we celebrate the dark winter with a midwinter festival. Bonfires, feasting and singing to the apple trees!

Chapter Four
Musonda Kabwe

I'm an illustrator from Johannesburg, South Africa. I enjoyed the project because it allowed me to spend Earth Day drawing whales, birds and mountains. During the pandemic, I spent a lot of time indoors, so I became very sensitive to how a few minutes in the sun can completely change my mood for the better.

Chapter Five
Cristina Merchán

I'm a freelance cartoonist, illustrator and graphic designer from Ecuador. I have been working independently under the name of Miti Miti since 2017. I live in Cuenca, a small city in southern Ecuador, where I also grew up. It is an honour for me as an Ecuadorian to have illustrated this chapter of the Inti Raymi festival in one of the most important temples of Tahuantinsuyo. I am interested in using my work to preserve and communicate the history, customs and traditions of my country.

Chapter Six & Front Cover
Nabila Adani

I'm a children's book illustrator from Jakarta, Indonesia. I briefly lived in Japan and the United States before coming back to my hometown. I enjoyed expressing the feeling of Pontianak through the colour of the sky in my artwork. I was also delighted to illustrate some mouth-watering local delicacies. As I grew up on the equator with constant sun all year long, I was excited and curious to read about other cultures in other parts of the world and how they celebrate the solstice.

Chapter Seven
Tinuke Fagborun

I'm a London-based, northern-born British-Nigerian Illustrator. I grew up in a Nigerian household with a dad who loved to tell me stories from West African folklore and Yoruba mythology. My style developed over time to incorporate my heritage, borrowing from its rich history, textiles and intricate patterns. Getting to work on this book was a dream for me as I got to use all the fun details and playful patterns that I love.

Chapter Eight
Ubahang Nembang

I am an illustrator, and sometimes a writer, from Nepal. I love to paint in watercolours, especially moody landscapes. In my stories I write about the interconnection between nature and humanity. Illustrating for *Solstice* has been a very enjoyable process. I hope that my artwork will give readers a glimpse of life in Nepal and that the variety of illustrators in the book will be a visual treat to all.

Chapter Nine
Sakina Saïdi

I am a French-Moroccan artist and illustrator living in London. I grew up in Lyon, France, in a Moroccan family. We would often spend summer in the imperial city of Fes, which is the centre for artisanship in Morocco. When I close my eyes, I can see the blue sky and hear the buzzing sounds of the Medina. Taking part in this project was a way to share the colours that filled my childhood.

Chapter Ten
QU Lan

I was born and raised in China. After graduating from the China Academy of Art, I moved to France to work as an illustrator and graphic designer. I collaborate with numerous major publishers and companies, and have won awards around the world.

Chapter Eleven
Mavisu Demirağ

Izmir, the city where I live and grew up, has hosted many civilisations from past to present. This is exactly why we have a great cultural diversity. Art, languages, sounds and tastes come together to form the cultural richness here. That's why I love this city. I also love being so close to the sea. In this project, I have been able to tell our stories through art. I think that sharing cultures is very valuable. I enjoyed taking part in this project, where art was our language allowing us to build bridges with colours.

Chapter Twelve
Vivian Mineker

I'm a freelance illustrator currently living in Amsterdam. I grew up in Taiwan and Portland, Oregon, U.S. I really enjoyed working on this project because I got to illustrate the Pacific Northwest where I spent a large part of my life. It's been really fun and nostalgic.

Chapter Thirteen
Gordy Wright

I'm an illustrator, printmaker and picture book maker from the UK and grew up in a small town on the edge of the North Yorkshire Moors. I later moved south to Bristol to study illustration at the University of the West of England. When I'm not painting and making books I enjoy being outside, be it riding my bike, having adventures in the countryside or just daydreaming in the garden.

Chapter Fourteen
Jannicke Hansen

My name is Jannicke and I'm an illustrator from the west coast of Norway. I love how the seasons change here. We have very dark winters, but when the Sun finally arrives, it really feels like we are seeing it for the first time all over again. This solstice I'm celebrating Sankthans with a big bonfire and lots of marshmallows!

Glossary

Ancestor A person from whom someone is descended, for example a grandparent.

Archaeologist Someone who studies prehistoric peoples and their cultures by looking at artifacts and remains.

Architect Someone who designs buildings.

Astrophysicist Someone who studies stars, planets, matter and space.

Atmosphere The mixture of gases that surround a planet.

Axis The imaginary line about which something, such as a planet, rotates.

Biracial Having a biological mother from one racial group and a biological father from another.

Botanist Someone who studies plants.

Breach To break through something, for example when a whale leaps up out of the water and breaks through the water's surface.

Chaps Leather leggings worn over a pair of trousers. Chaps are part of many South American traditional outfits.

Dawn The first appearance of daylight in the morning.

Drenched Completely wet; soaked.

Durian Fruit with a hard, prickly rind, pulpy flesh and a strong odour that some people find unpleasant.

Eerie Uncanny; spooky; weird.

Equinox When the Sun crosses Earth's equator, making night and day approximately equal lengths all over Earth. This occurs around the 21st of March and the 22nd of September.

Evaporate Become vapour, for example when water is heated to boiling point.

Fillet A boneless cut or slice of meat or fish; the act of cutting off a fillet.

Flare A device used to create a blaze of light.

Generator A machine that creates power, often electricity.

Geometric A regular, repeated pattern.

Glaze A special kind of paint that makes ceramics glossy.

Harbour A port with docks for boats that keeps them sheltered from winds.

Hemisphere Half of a planet, for example, Earth is divided between northern and southern hemispheres along the equator.

Hippodrome An arena built for horse races and chariot races.

Horizon The line that separates Earth and the sky.

Humid Containing a high amount of water or water vapour. The air is often described as humid when it feels noticeably moist.

Impediment An obstruction; something in the way.

Isolated Separate from other people or things; alone.

Magnetic field The area around a magnet in which its force can be felt. Earth has a magnetic field that is created by the movement of its molten iron core.

Migration To travel from one country, region or place to another.

Mortar and pestle A bowl and tool used for grinding things, such as spices.

Mosque A Muslim temple.

Obelisk A tall stone monument with four sides and a pyramid on top.

Pageant A show that often includes a procession of people in costumes.

Phenomenon Something that is special or extraordinary.

Pollinate To move pollen from one flower to another so that the plant can produce fruit.

Poncho A blanket-like cloak with a hole in the centre for a person's head to go through.

Poverty The state or condition of having little or no money, goods or means of support.

Preserve A way of storing something so that it lasts a long time. For example, vegetables can be jarred or canned so that they can be eaten months after they have been picked.

Professor A teacher, usually in a university.

Propane A kind of gas fuel.

Prosperity The state of having wealth and good fortune.

Ritual A set of words and actions that are repeated on particular occasions, often religious.

Rugged Rough; harsh; rocky.

Scaffolding A temporary structure for holding workers and materials while a building is built or repaired.

Skipper Captain of a ship.

Smog Unhealthy or irritating air pollution.

Solar From the Sun, for example solar power.

Solstice Two times of year when the Sun's path is furthest north or south from the equator. This happens around the 21st of June and around the 22nd of December. At the winter solstice the length of the day is the shortest in the year and at the summer solstice the day is the longest in the year.

Spawning ground A place where fish, amphibians, molluscs or crustaceans deposit masses of eggs at once.

Spiritual Relating to the spirit or soul, as distinguished from physical nature.

Strait A narrow passage of water connecting two large bodies of water.

Tango Music for a Latin American dance of the same name.

Tradition The handing down of statements, beliefs, legends, customs, information, stories, rituals and practices.

Twilight The time between daybreak and sunrise or sunset and nightfall when the light is soft and the Sun is below the horizon.

Vault A secure, underground building.

Vendor Someone who sells things, such as a street-food vendor.

Warren A system of tunnels or corridors.

Index

A
Amazigh people 39
Amundsen-Scott Research Station (Antarctica) 4–7
Andes Mountains 8
Annapurna (Nepal) 32
Antarctica 4–7
Arabic 39
archaeology 53
architecture 45, 46
Arctic Circle 56
Argentina 8–11
arts festivals 12–15
astrophysics 6
Atlantic Ocean 8
Aurora Australis 7
Australia 12–15

B
bamboo 41
bāo zi 41
Barage (Nigeria) 28–31
batteries 30
Beagle Channel 11
Beijing (China) 40–3
Big Bang 6
birds 18, 38, 50
blackout curtains 57
bonfires 10, 13, 58
Borneo (Indonesia) 24–7
Bosphorus (Turkey) 46
Bunny Chow 19
Burning of Obstacles and Impediments 10

C
Cape Town (South Africa) 16–19
cats, stray 44, 45, 47
çaydanlık 44
ceramics 37
China 40–3
chiya 33, 35
coal mining 57
Cuzco (Peru) 22

D
Darija 39
dawn 55
diving 17
Dragon Boat Festival 42–3
Druids 55

dry season 25, 29
durians 27

E
Earth
  axial tilt 2, 4, 24, 29
  magnetic field 7
Ecuador 20–3
equator 2, 20, 24
equinoxes 29
evil spirits 43, 58

F
festivals 12–15, 21–3, 42–3, 58
fireworks 11
fishing industry 48–51

G
Global Seed Vault (Svalbard) 57
Gurkas 33

H
Hagia Sofia (Istanbul) 45–6
hemispheres 2
henges 52
Hobart (Tasmania) 12–15
humidity 25, 40
hydroponics 5

I
ice skating 10
Ilwaco (Washington, USA) 48–51
Incas 20–2
Indonesia 15, 24–7
Ingapirca (Ecuador) 20–3
Inti Raymi 20–2
Istanbul (Turkey) 44–7

J
Japan 4, 14

L
literacy 33
Longest Night 9–11
Longyearbyen (Svalbard) 56–9

M
March of Light 11
Marrakesh (Morocco) 36–9
marriage, interracial 19

midsummer traditions 58
midwinter 12, 60
momos 33, 35
Morocco 36–9
mosques 45–6
music 11, 13, 22

N
Nepal 32–5
Nigeria 28–31
North Pole 2, 4
Northern Hemisphere 2
Norway 2, 56–9

O
ogoh-ogoh parade 15
ornithology 18

P
Pachamama 21–2
Pacific Ocean 8
pagans 55
pandas 41
Parque Nacional Tierra del Fuego 9
parties 9–11, 29
pigeons 38
plants 5, 25, 43, 51, 57, 58
Pokhara (Nepal) 32–5
polar bears 57
pollination 5
Pontianak (West Kalimantan, Borneo) 24–7
Port St Johns (South Africa) 16–19

Q
Quechua people 22
Qur'an 26

R
rice farming 25

S
sacrifices 22
salmon 50–1
Sapa Inca 22
sardine migration 16–17
school 35, 39, 40, 41–2, 46, 51
seasons 2, 25, 29
seeds 57
shade 30
Shalako 14

skiing 9
smog 40
solar power 30
solstices 2, 29, 56
  summer 20, 24, 29, 32, 37, 39, 42, 44, 49, 52–3, 55, 56
  winter 4, 8, 12, 16, 20, 24
Souk Semmarine (Marrakesh) 36–7
South Africa 16–19
South Pole 2, 4–7
Southern Hemisphere 2
sports 12, 26, 47
St John's Eve 58
stone circles 52–5
Stonehenge (UK) 52–5
stonemasons 45
sunlight 2, 4, 8, 12, 16, 20, 24, 28, 30, 32, 36, 40, 44, 48, 52, 56
sunsets 49
Svalbard (Norway) 56–9

T
tagines 37, 38
Tamazight 39
Tasmania (Australia) 12–15
Tata Inti 21, 22
temples 22
Tierra del Fuego (Argentina) 8–11
Turkey 44–7

U
United Kingdom 52–5
United States of America 48–51
universe 6
Ushuaia (Argentina) 8–11

W
water sports 26
wet season 25, 29, 32, 33
whales 17
Wiccans 55
wildflowers 58
Winter Feast 13

Z
Zuni people 14

Source Notes

Here is a selection of the books and articles that the author used as reference material for this book.

Algarra, Alejandro. *What Causes Weather and Seasons?* (Barron's, 2016)

Asael, Anthony. *Children of the World* (Universe Publishing, 2011)

Atinuke. *Africa, Amazing Africa: Country By Country* (Candlewick, 2021)

Bills, Garland D. *An Introduction to Spoken Quechua* (University of Texas Press, 1969)

Blexbolex. *Seasons* Translated by Claudia Bedrick (Enchanted Lion Books, 2009)

Bowden, Rob. *Global Cities: Cape Town* (Chelsea House, 2007)

Bowden, Rob. *Global Cities: Istanbul* (Chelsea House, 2006)

Bullock, Marita. *The Big Book of Festivals* (Lothian Children's Books, 2021)

Butterfield, Moira. *Food Around the World* (Cavendish Square, 2016)

Cascadia Department of Bioregion. 'Chinook Wawa' cascadiabioregion.org/chinook-wawa

Coan, Sharon. *What the Sun Can Do* (Time for Kids, 2015)

Cooper, Susan. *The Shortest Day* (Candlewick, 2019)

Corr, Christopher and Claire Grace. *A Year of Celebrations and Festivals* (Frances Lincoln Children's Books, 2021)

D'Aluisio, Faith. *What the World Eats* (Tricycle Press, 2008)

Echols, John M. *An English-Indonesian Dictionary* (Cornell University Press, 1990)

Galat, Joan Marie. *Cultural Traditions in Turkey* (Crabtree, 2016)

Gibbons, Gail. *The Reason for the Seasons* (Holiday House, 1995)

Handelsman, Michael. *Culture and Customs in Ecuador* (Greenwood Press, 2000)

Iz, Fahir. *Oxford Turkish Dictionary* (Oxford University Press, 2002)

Jackson, Ellen B. *The Summer Solstice* (Millbrook, 2001)

Jackson, Ellen B. *The Winter Solstice* (Millbrook, 1997)

Kirkeby, Willy. *English-Norwegian Dictionary* (Norwegian University Press, 1990)

Laroche, Giles. *If You Lived Here: Houses of the World* (Houghton Mifflin, 2011)

Lonely Planet: *Indonesia* (Lonely Planet, 2020)

Lonely Planet: *Morocco* (Lonely Planet, 2021)

Lonely Planet: *Nepal* (Lonely Planet, 2020)

Malerba, Giulia. *Food Atlas: Discover All the Delicious Foods of the World* (Firefly Books, 2017)

Markle, Sandra. *Super Cool Science: South Pole Research Stations Past, Present and Future* (Walker, 1998)

McGinty, Alice B. *Feasts and Festivals Around the World: From Lunar New Year to Christmas* (Little Bee Books, 2021)

McGinty, Alive B. *From Pancakes to Parathas: Breakfast Around the World* (Little Bee Books, 2019)

Morganelli, Adrianna. *Cultural Traditions in Argentina* (Crabtree, 2016)

Murphy, Charles. *Food Around the World* (Gareth Stevens Publishing, 2017)

National Science Foundation. 'Amundsen-Scott South Pole Station' www.nsf.gov/geo/opp/support/southp.jsp

Newman, Roxana Ma. *An English-Hausa Dictionary* (Yale University Press, 1990)

Onyefulu, Ifeoma. *One Big Family: Sharing Life in an African Village* (Frances Lincoln, 2006)

Oregon Encyclopedia 'Chinook Jargon (Chinook Wawa)' www.oregonencyclopedia.org/articles/chinook_jargon/

Owings, Lisa. *Stonehenge* (Bellwether, 2015)

Oxford Arabic Dictionary (Oxford University Press, 2018)

Oxford Chinese Dictionary (Oxford University Press, 2010)

Oxford Japanese Dictionary (Oxford University Press, 1993)

Oxford Spanish Dictionary (Oxford University Press, 2008)

Pellegrini, Nancy. *Global Cities: Beijing* (Chelsea House, 2008)

Peppas, Lynn. *Cultural Traditions in China* (Crabtree, 2016)

Perritano, John. *Australia: Tradition, Culture and Daily Life* (Mason Crest, 2015)

Perritano, John. *China: Traditions, Culture and Daily Life* (Mason Crest, 2015)

Perritano, John. *South Africa: Traditions, Culture and Daily Life* (Mason Crest, 2015)

Pfeiffer, Wendy. *The Longest Day: Celebrating the Summer Solstice* (Dutton, 2010)

Pfeiffer, Wendy. *The Shortest Day: Celebrating the Winter Solstice* (Dutton, 2003)

Raj, Prakash A. *Nepali-English English-Nepali Dictionary and Phrasebook* (Hippocrene Books, 2002)

Raum, Elizabeth. *Stonehenge* (Amicus, 2014)

Stewart, Whitney. *What's On Your Plate? Exploring the World of Food* (Sterling, 2018)

Storad, Conrad J. *Earth is Tilting!* (Rourke, 2011)

Teckentrup, Britta. *Look at the Weather* (Owlkids, 2018)

Terp, Gail. *What is a Solstice?* (The Child's World, 2016)

The Cities Book (Lonely Planet Kids, 2016)

Walrond, Beth. *A Taste of the World: What People Eat and How They Celebrate Around the Globe* (Little Gestalten, 2019)

Webb, Mick. *The Book of Languages: Talk Your Way Around the World* (Owlkids, 2015)

Zoehfeld, Kathleen Weidner. *Secret of the Seasons: Orbiting the Sun in Our Backyard* (Random House Children's Books, 2014)